Solve It 2
Manipulative Math Puzzles

by
Dr. Alan Barson

Good Apple

Editor
Susan Eddy

Designer
Ruth Otey

Cover
Peter Van Ryzin

GOOD APPLE
A Division of Frank Schaffer Publications, Inc.
23740 Hawthorne Boulevard
Torrance, CA 90505-5927

4 5 6 7 8 9 MAL 01

CONTENTS

INTRODUCTION

Solve It 2 contains 24 manipulative math puzzles for grades 6–8 that conform to the definition of problem-solving tasks as set forth by the National Council of Teachers of Mathematics (NCTM). Each activity asks students to use cards numbered 1–9 as manipulative devices to solve problems on a task sheet. The NCTM standards that are addressed in *Solve It 2* include

- selecting mathematical tasks to engage students' interests and intellects

- providing opportunities to deepen students' understanding of the mathematics being studied and its applications

- orchestrating classroom discourse in ways that promote the investigation and growth of mathematical ideas

- using, and helping students use, technology and other tools to pursue mathematical investigations

- guiding individual, small-group, and whole-class work

Task sheets are designed to encourage students to explore mathematical concepts such as place value, large numbers, estimations, rounding, whole-number equations, equivalency, and problem-solving strategies. The use of number-card manipulatives heightens student interest and eliminates the frustration of erasures and cross-outs while in search of puzzle solutions. For example, when asked to find solutions using the nine number cards to make this addition problem work, students are able to effectively find the pattern and the more than 200 solutions by manipulating the cards.

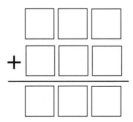

Solve It 2 is divided into four sections: Games for One, Games for Two, Algorithm or Equation Activities, and Problem-Solving Situations. In Games for One, students search for patterns and compete against themselves, using the fewest possible moves to find answers. In Games for Two, students compete with partners to find solutions and win the games. Strategies utilized in these games include working backwards, finding patterns, using logic, reducing to a simpler pattern, and following the process of elimination. The activities in Equation or Algorithm Activities involve understanding place value and number operations. Problem-Solving Situations is a potpourri of activities using logic and patterns to solve a series of interesting tasks.

The activities in *Solve It 2* work equally well as meaningful drill-and-review exercises, lesson introductions, homework assignments, or learning-center and cooperative-group activities. You may wish to challenge advanced students to design puzzle activities of their own that use one or more sets of the nine number cards.

The number cards (found on the inside back cover) may either be cut out from the cover itself, or reproduced and laminated for durability. Keep sets of number cards in envelopes or plastic zipper bags and store with the gameboards for easy accessibility.

SWITCH!

Place number cards 1–8 in the appropriate spaces below. The object is to switch the odd numbers with the even numbers in the fewest moves. The numbers do not have to be in order. You can do this only by moving a card to an empty space or by jumping over one opposite card (odd over even or even over odd) to an empty space.

Odd	Odd	Odd	Odd	Switch!	Even	Even	Even	Even

BOX PROBLEM

Place number cards 1–9 in the empty grid as shown in Example 1. Remove number card 9. The object is to rearrange the numbers into the arrangement shown in Example 2, using the fewest possible moves. You may only slide a number into an adjacent empty space. No jumping is allowed. This can be done in 22 moves.

Example 1

7	5	6
8	3	2
4	9	1

Example 2

1	2	3
4	5	6
7	8	

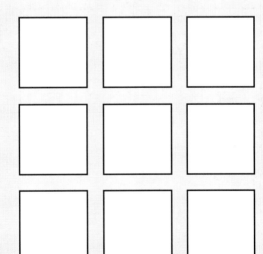

JUMP SUMS

Place number cards 1–9 on their corresponding squares. The object is to jump over one number into an empty space and remove the number you jump over until you can't make anymore jumps. Follow the lines when jumping. For example, you might start by jumping 3 over 1 or 5 over 2. You may only jump over one number at a time. The empty space must be next to the number you jump. Your score is the sum of the numbers you capture on the jumps.

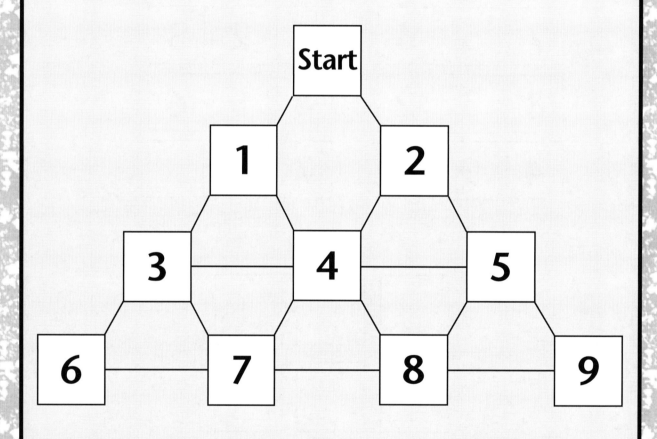

FOUR MOVES ONLY

Place the number cards on the appropriate squares. The object is to rearrange the numbers into an odd-even-odd-even sequence in just four moves. You may only pick up two adjacent cards with each move. It is permissable to move the number cards outside of the row of squares (either to the left or the right). Thus, in the final sequence, some of the number cards may be positioned either to the left or right of the row of squares. There may also be blank squares between some of the numbers in the final sequence.

1	3	5	7	2	4	6	8

MASTER CHALLENGE
Can you do it in three moves?

TRIANGULAR TOTALS

Place number cards 1–9 face down in a pile on the square below. Take the top card and place it face up anywhere on the grid. Continue until all the cards have been placed. Then add across and down and write the sums in the triangles. Your score is the total of all the sums that appear more than once in the triangles. See if you can get a score greater than 70. Can you determine the highest possible score?

HIGH OR LOW

Before beginning, decide whether the high score or the low score will win the game. Get a piece of scrap paper. Place number cards 1–9 face down on the grid, one card per space. On each turn, choose two fully adjacent cards (sides touching, not corners) and either multiply the two numbers or subtract the smaller from the larger. Write down the answer. When all cards have been removed but one, players total their scores. The last card is turned over and both totals are divided by the number on the last card. Play the game three times and total the three scores. High total or low total wins—whichever was decided prior to beginning play.

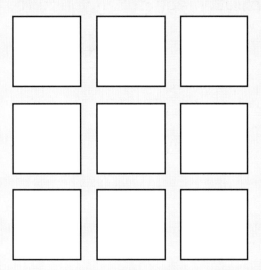

35

Shuffle the number cards 1–9 and place them face down in a pile on the square below. The object of the game is to get your score as close as possible to 35 without going over.

Start by turning over the top card. Choose one of the math operations and turn over the next card. Do the computation, choose another operation, and turn over the next top card. You can stop at any time, as long as you don't exceed 35. Your opponent then tries to get closer to 35 than you did without going over.

You may use only the following five operations. You *must* use the multiplication sign during your turn.

$$+ \quad + \quad - \quad X \quad +$$

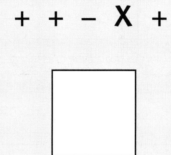

OPERATION

Shuffle the number cards 1–9 and place them in a pile face down on the square below. The object is to be the player closest to 25 without going under 25.

Turn over the top card and choose an operation from the list below. As you use the operations, cross them out. Turn over the next card, do the computation using the operation you chose, and subtract that total from 100. Continue in this manner as long as you wish. You may stop any time and end your turn. For example:

         Top card: 8
         Operation chosen: +
         Next card: 5
         Computation total: 13
         100 – 13 = 87
         Next card: 6
         Operation chosen: X
         Next card: 9
         Computation total: 54
         87 – 54 = 33
         Stop.

Operations:
+ + – – X

MULTIPLICATION STARS

Place number cards 1–9 on the squares provided. Each player will need three coins or markers. The object is to get three markers in a row horizontally, vertically, or diagonally, on the scoreboard below.

To start, one player removes one of the cards and places it on the star. The other player takes a card and multiplies that number by the number on the star. If the player matches a number on the scoreboard, he or she covers that number with a marker. The next player then multiplies the number on the star by a number he or she chooses from the squares. If the answer is the same as one covered by the other player, the marker is replaced by the second player's marker. The game continues until all the cards are used (tie) or until one player gets three markers in a row on the scoreboard.

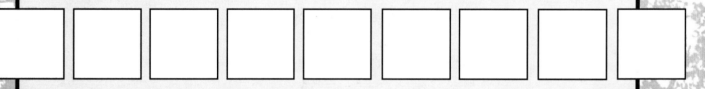

8	9	15
24	12	21
6	32	27

Games for Two

THE GRID GAME

Shuffle number cards 1–9 and place them face down in a pile on the square below. The first player turns over the top card and places it anywhere in Grid 1. The other player turns over the next card and places it anywhere in Grid 2. The game continues in this manner until one card is left. When this is turned over, both players must use it in the remaining space in their grids. Players complete the computation to see who has the highest score. Players may wish to try using each of the different sets of grids once (see Grids 3, 4, 5, and 6 on page 16) and totaling their three scores.

Grid 1

Grid 2

Grid 4

Grid 3

Grid 6

+

Grid 5

+

52

Place number cards 1–9 in the squares so that the answer works out to 52.

$$\square\ \square\ +\ \square\ \square\ +\ \frac{\square\ \square}{\square\ \square}\ +\ \square\ 0\ =\ \square\ \square$$

TELLING THE TRUTH

Place number cards 1–9 in the squares in ascending order. Place two subtraction signs and one addition sign within the number row to make it a true sentence.

☐ ☐ ☐ ☐ ☐ ☐ ☐ ☐ ☐ = 100

MAKING 100

Place number cards 1–9 in the boxes below along with three + signs and two = signs to make an equation totaling exactly 100. Adjacent number cards may represent two-digit numbers, if necessary (for example, 1 9 + 6 + 4 + 3 5).

PATTERN SEARCH

Place number cards 1–9 in the addition problem below. How many ways can you find to make it true? There are more than 200 if you find the pattern!

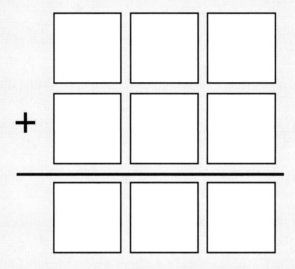

ALGORITHM

Place number cards 1–9 in the grids below to obtain the requested results.

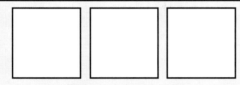

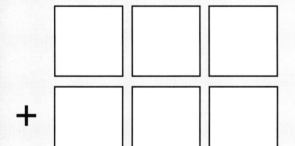

Largest Possible Sum

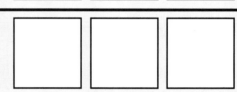

Smallest Possible Sum

As Close as Possible to 500

As Close as Possible to 800

Algorithm or Equation Activities

KEEP IT PROPER

If you add up the odd-numbered cards, the sum is 25. The even-numbered cards add up to 20. Can you use the grid below to arrange the number cards so that both the odd- and even-numbered digits add up to the same total? Improper fractions and recurring decimals may not be used.

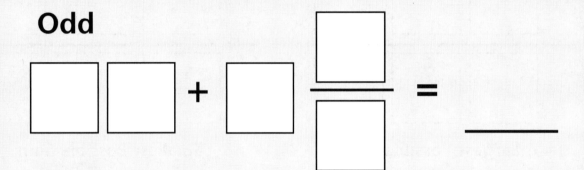

Odd

$$\Box\Box \ + \ \Box\ \dfrac{\Box}{\Box} \ = \ \underline{\quad\quad}$$

Even

$$\Box\Box \ + \ \dfrac{\Box}{\Box} \ = \ \underline{\quad\quad}$$

Algorithm or Equation Activities

EVEN TREATMENT

Place number cards 1–9 in the squares below in such a way that all the totals in the circles are even numbers. Compute the problems by row (left to right) and column (top to bottom) and write the answers in the circles.

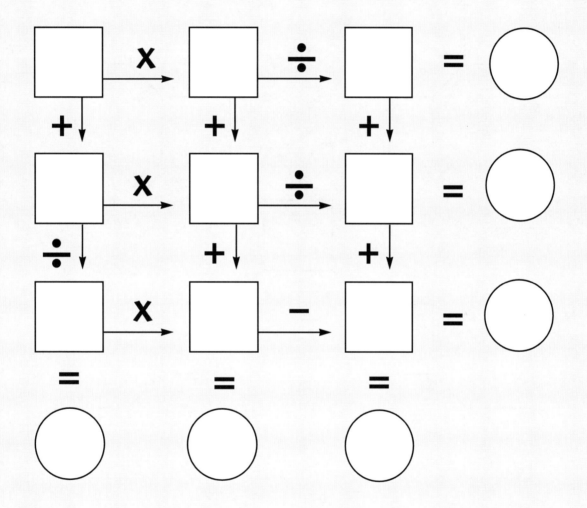

Algorithm or Equation Activities

SLANTS AND CORNERS

Place number cards 1–9 in the squares below so that the sum of each diagonal line is 25 and the sum of the four corner numbers is also 25.

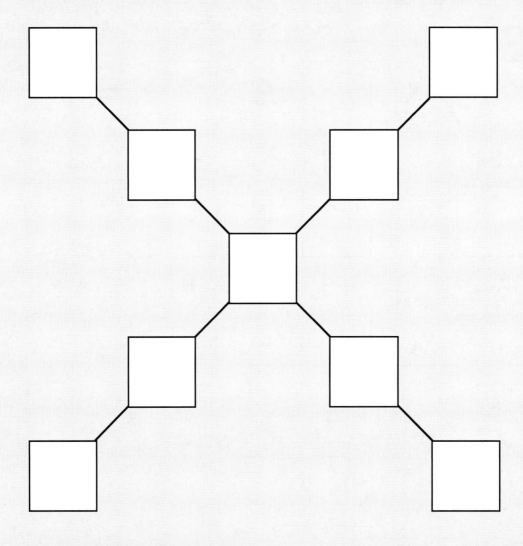

Problem-Solving Situations

© 1997 Good Apple

SQUARE DANCE

Can you arrange number cards 1–9 so they form four square numbers?

MASTER CHALLENGE

The master challenge is to arrange the number cards into one number that is a nine-digit square number!

MYSTERY SQUARE

Follow these directions to place number cards 1–9 on the grid below.

1. The corner cards are all factors of 24.
2. The middle row contains consecutive odd numbers but not in consecutive order.
3. The two diagonals add up to the same number.
4. The first column and the middle row have the same total.
5. The opposite corners add up to the same total.
6. The middle column has the highest total—six more than the right-hand column.

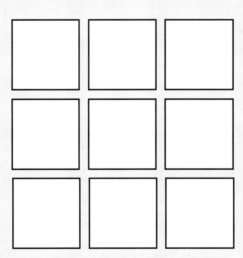

DON'T GET CROSS

Can you place number cards 1–9 in each cross so that the sum of the numbers in the horizontal arm of the cross equals the sum of the vertical arm? There are many different ways to do each one.

ONCE, TWICE, THRICE

Place number cards 1–9 on the grid below so that the three-digit number in the second row is twice the one in the top row, and the three-digit number in the bottom row is three times the one in the top row. Can you find all four ways to solve the problem?

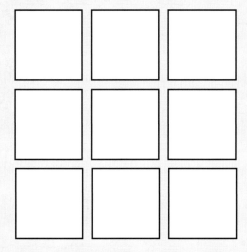

X MARKS THE SPOT

Place number cards 1–9 in the grid below so that the sum of each arm totals 23. Then try to get each arm to total 24, 25, 26, and 27.

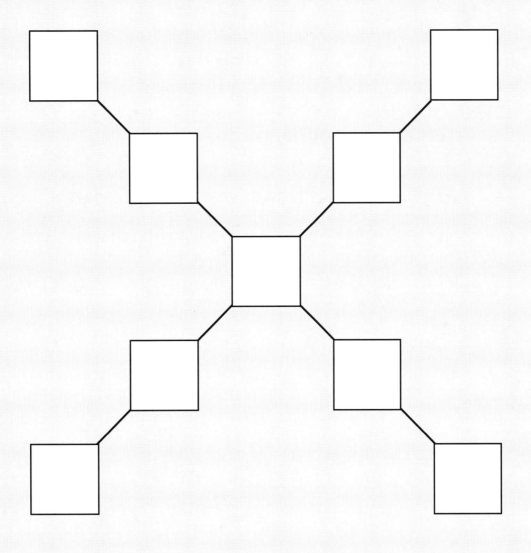

FAVORITE NUMBERS

Use the clues below to find each person's favorite number. Place the appropriate number cards below each person's name.

1. When you square Dobs's number and add the digits, it equals itself.
2. Debbie's number is Dobs's number turned upside down.
3. Lisa's number is prime and the digits are consecutive.
4. Bernadette's number is evenly divisible by 3, 5, and 25.
5. Ruth's number is evenly divisible by 2, 4, and 8, and the digits are in ascending order.

Lisa

☐ ☐

Dobs

☐

Bernadette

☐ ☐

Debbie

☐

Ruth

☐ ☐ ☐

ANSWER KEY

Switch! page 6
The secret is to keep the numbers alternating. Don't get caught with two odds or evens together. Also try to do the problem in pairs. Start with one pair, then two pairs, then three pairs, and so on. The minimum number of moves for each set of pairs is

Pairs	Moves
1	3
2	8
3	15
4	24

Box Problem, page 7
L= Left D= Down R=Right U=Up
1L, 2D, 6D, 5R, 3U, 1U, 2L, 6D, 5D, 3R, 1U, 2U, 4R, 8D, 1L, 2U, 4U, 8R, 7D, 4L, 5L, and 6U

Four Moves Only, page 9
1. Move 7 and 2 to outer left.
2. Put 6 and 8 to the empty spaces from 7 and 2.
3. Put 5 and 6 at right end.
4. Put 3 and 8 at right end.
Result: 7 2 1 4 5 6 3 8

Master Challenge (3 moves)
1. Move 7 and 2 to outer left.
2. Move 5 and 4 to left of 7 and 2.
3. Move 3 and 6 to outer right.
Result: 5 4 7 2 1 8 3 6

52, page 17
$49 + \underline{76} + 1 + 0 = 52$
38

Telling the Truth, page 18
$123 - 45 - 67 + 89 = 100$

Making 100, page 19
$15 + 36 + 47 = 98 + 2 = 100$

Pattern Search, page 20
The secret is to realize that all the answers total 18 and you must carry once in the ones or tens place. You can make all the totals of 18 except those that have a two in the hundreds place.
$$\begin{array}{r} 4\;6\;7 \\ \underline{3\;5\;2} \\ 8\;1\;9 = 18 \end{array}$$

Algorithm, page 21

Largest	Smallest
2 4 6	2 7 6
7 3 5	1 8 3
9 8 1	4 5 9

Closest to 500	Closest to 800
1 6 7	6 5 8
3 2 8	1 3 4
4 9 5	7 9 2

Keep It Proper, page 22
$79 + 5\,1/3 = 84\,1/3$
$84 + 2/6 = 84\,1/3$

Even Treatment, page 23
$8 \times 2 \div 4 = 4$
$+ \quad + \quad +$
$6 \times 9 \div 3 = 18$
$\div \quad + \quad +$
$7 \times 5 - 1 = 34$
$= \quad = \quad =$
$2 \quad 16 \quad 8$

Slants and Corners, page 24

```
9           8
   3    1
      7
   6    2
4           5
```

Square Dance, page 25

9, 81, 324, 576

Master Challenge

139, 854, 276 = square of 11,826

Mystery Square, page 26

```
2 7 4
1 5 3
6 9 8
```

Don't Get Cross, page 27

```
   2              1
4 5 1  6 7        2
   3           3 4 9 5 6
   8              7
   9              8
```

Once, Twice, Thrice, page 28

```
1 9 2      2 1 9
3 8 4      4 3 8
5 7 6      6 5 7

2 7 3      3 2 7
5 4 6      6 5 4
8 1 9      9 8 1
```

X Marks the Spot, page 29

```
5       2       1       6
  6  3     7  5
     1  = (23)     3  4 = (24)
  9  4     8  4
8       7       2       9

1       2       9       1
  3  6     3  8
     5 9 = (25)     7  2 = (26)
  4  9     4  2
8       7       6       5

        8       6
          2   1
             9   5 = (27)
          4   5
        7       3
```

Favorite Numbers, page 30

Dobs = 9, Debbie = 6, Lisa = 23,
Bernadette = 75, Ruth = 128